Dedicated to our kids, Noor and Ettore.
To Marisa, Carina, and Vanessa.
And to all the children who keep shaping
the future of the world. We hope they are always
motivated to explore the path of science
and contribute to a better tomorrow.

A book for children to understand the power of Hydrogen

Little great Hydrogen

Andrea Menegazzo
Magdalena Cochanski Rodríguez

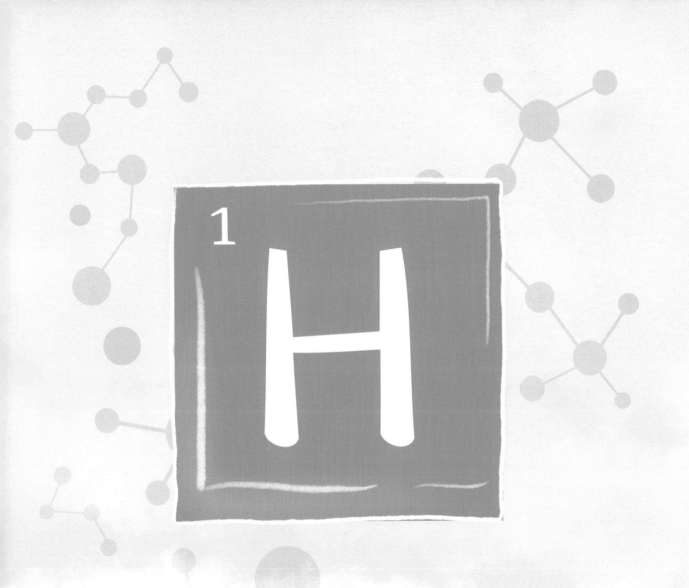

I am the simplest and lightest element.

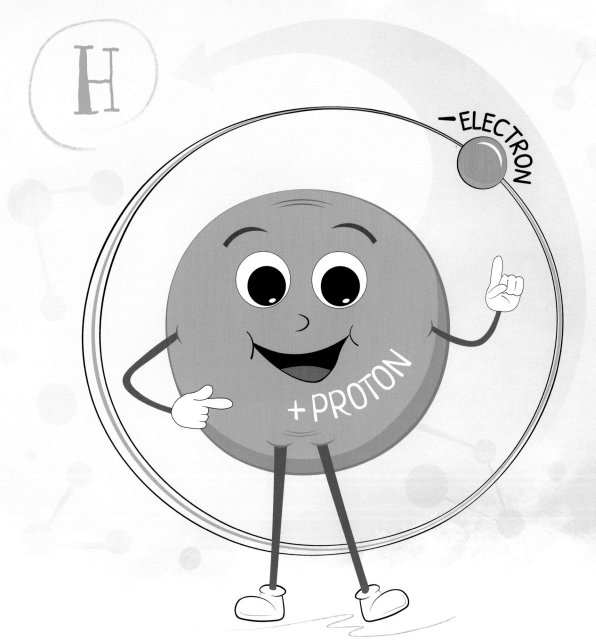

There are over 100 elements in the world.

H																	He
Li	Be											B	C	N	O	F	Ne
Na	Mg											Al	Si	P	S	Cl	Ar
K	Ca	Sc	Ti	V	Cr	Mn	Fe	Co	Ni	Cu	Zn	Ga	Ge	As	Se	Br	Kr
Rb	Sr	Y	Zr	Nb	Mo	Tc	Ru	Rh	Pd	Ag	Cd	In	Sn	Sb	Te	I	Xe
Cs	Ba	La	Hf	Ta	W	Re	Ir	Pt	Au	Au	Hg	Ti	Pb	Bi	Po	At	Rn
Fr	Ra	Ac	Rf	Db	Sg	Bh	Hs	Mt	Ds	Rg	Cn	Nh	Fl	Mc	Lv	Ts	Og

		Ce	Pr	Nd	Pm	Sm	Eu	Gd	Tb	Dy	Ho	Er	Tm	Yb	Lu

Some of them are:

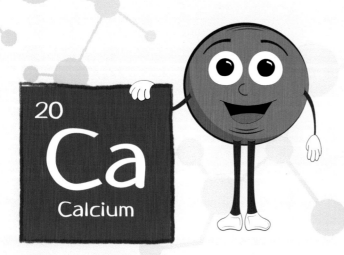

All the things around us are made up of elements.

We are in the sun,

in the air,

and even in your shoes.

We are everywhere!

I don't like being alone.
I am always together with another Hydrogen atom.

Together we make a molecule H_2.

We are present as a gas with no color, no smell...

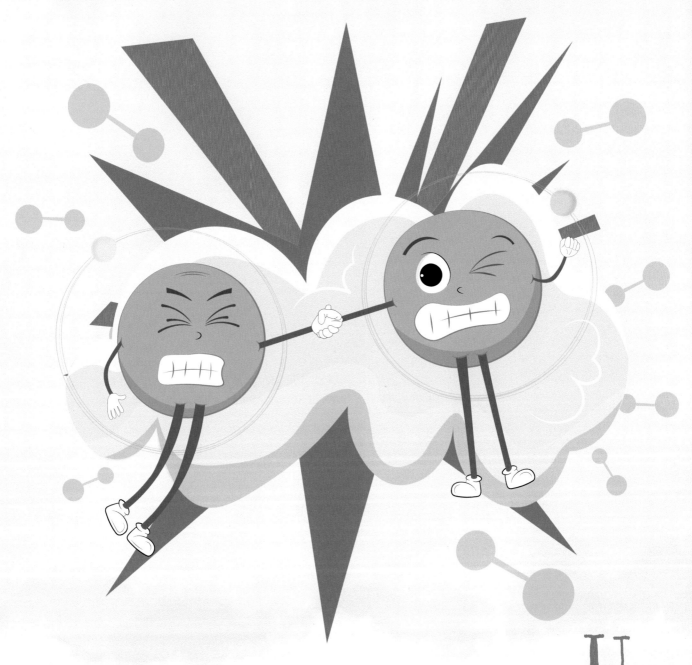

We are very energetic when we get together as H$_2$.

We become a fuel that can power homes,

cars,

trucks,

even rockets!

WE CAN POWER EVERYTHING!

But you won't find H_2 molecules hanging around without company.

We usually hang out with other elements, and together we make a lot of things.

For example, when we get together with Oxygen, we make water.

H_2O

If the sun or wind gives the energy to separate us from water, then we'll be called:

You can get H_2 from other sources too, like natural gas. But you might also get a lot of our CO_2 friend in the process,

NATURAL GAS

CARBON DIOXIDE
CO_2

The temperature will rise if CO_2 is in the air.

And we don't want the world to be hotter.

We want it to be comfortable and pleasant.

Scientists keep experimenting to find better ways to separate me and my other Hydrogen friends from other elements,

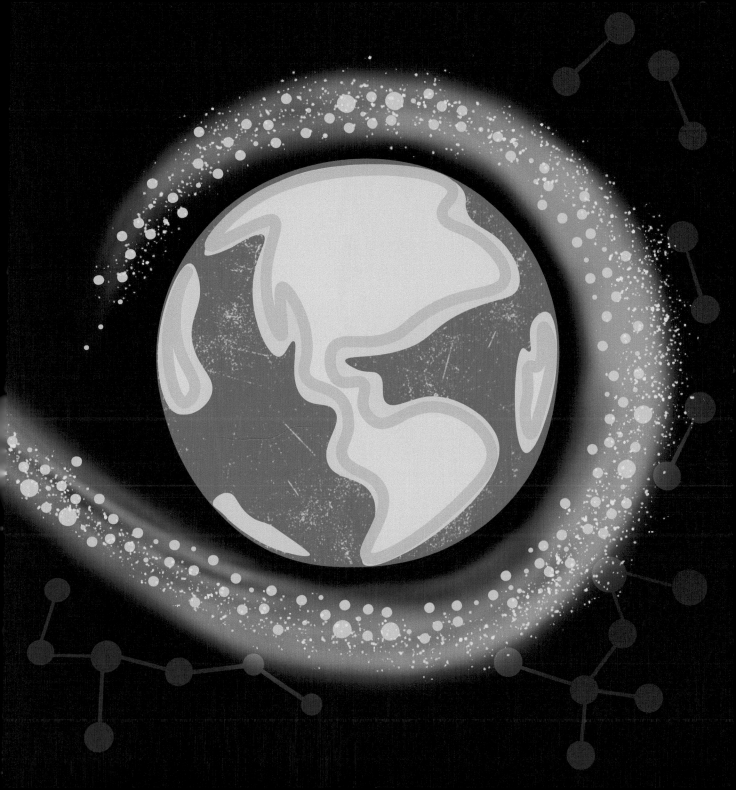

Andrea Menegazzo is a consultant focused on chemicals, energy, and industrials. He is passionate about fostering scientific progress. As a father, he started to write books to inspire his children to explore the wonders of science. Andrea is originally from Italy and now lives in the USA with his wife and children.

Magdalena Cochanski Rodríguez writes and illustrates for children. With a background in architecture, she has been creating designs and architectural visuals before becoming an entrepreneur to provide learning tools.
She founded *hola hallo* to offer bilingual books for children. You can learn more at www.hola-hallo.de. She was born and raised in Mexico, lived with her husband and children in the USA, and now resides in Germany.

Text and Illustrations © 2023 Andrea Menegazzo and Magdalena Cochanski Rodríguez

All rights reserved.
No part of this publication may be reproduced in whole or in part, or stored in a retrieval system, or transmitted in any form or by any means, electronic, mechanical, photocopying, recording, or otherwise, without permission of the author or publisher.

ISBN: 979-8-9883807-0-2

Written by Andrea Menegazzo and Magdalena Cochanski Rodríguez
Illustrated by Magdalena Cochanski Rodríguez

MOGA BOOKS

www.mogabooks.com